Anonymous

A List of the Plants in Botanic Garden

Anonymous

A List of the Plants in Botanic Garden

ISBN/EAN: 9783337337193

Printed in Europe, USA, Canada, Australia, Japan

Cover: Foto ©berggeist007 / pixelio.de

More available books at **www.hansebooks.com**

A List of the Plants in Botanic Garden

Section A.

Clematis — — from Dr. Gray.
" sp. New Mexico. Dr Thurber.
" Catesbyana
" sp. (Regel).
" longerina
" orientalis (Regel March 2 1878)
" sp.
" ligusticifolia (Oregon.
" sp.
" paniculata (Japan.
" sp. (")
" sp. (" .
" grata. (Himalaya Mts.)
" vitalba (Europe.)
" campaniflora (Portugal.)
" sp.
" sp.
" Hendersoni.
" sp.
" sp.
" viticella var purpurea (S. Europe)
" sp.
" sp.
" sp.
" coerulea (Japan).
" sp.
" apiifolia. (Japan)
" Pitcheri (W. States)
" sp.
Juglans cinerea (Butternut) (U.S.)
Menispermum Canadense (Moonseed) (U.S).
Aesculus flava. var purpurascens Purple Buckeye (U
Actinidia polygama (Eastern Asia)
Celastrus orixa (Japan)
" scandens. Waxwork. (U.S.)
" punctatus (Japan)
Morus alba. (White Mulberry) (China ;
Fraxinus excelsior. Eur. Ash.
Vitis indivisa U. States.
Crataegus tomentosa var mollis (W. States)

Bed No 1.

Clematis integrifolia (Europe)
" recta (")
" Davidiana (N. China)
" tubulosa (")
" crispa (Woolson 1878)
" ochroleuca. (Mid states & South)
Thalictrum purpurascens (U.S.)
" simplex
" cornuti (U.S)
" dioicum Early Meadow Rue. (U.S)
" minus (Europe)
" tuberosum (")
" elatum (")
" angustifolium (")
" glaucum (S. Europe)
" Fendleri (Rky. Mts.
Anemone cylindrica (U.S.)
" Virginica (")

Bed No 2.

Anemone Japonica. (Japan)
" " var Honorine Jobert. (Hort)
" Pulsatilla (Europe)
" pratensis (")
" sylvestris (")
" patens var Nuttalliana (W. States)
" Pennsylvanica (U.S.)
Hepatica acutiloba (")
" triloba (")

1

Bed No 3 Continued

Delphinium magnificum (March 25 1878. Thompson)
" moschatum
" cashmerianum (Leichtlin Feb 11. 1878)
" exaltatum (S. States)
Trollius Asiaticus (Siberia).
" napellifolius
" Europaeus (Europe) (Globe flower)
Helleborus orientalis (Greece) (Hellebore)
" atrorubens (Europe,
Isopyrum biternatum (W. States)
" thalictroides (Europe)

Bed No 4

Nigella Damascena (Europe)
" Hispanica (Spain)
" orientalis Levant.
Aquilegia glandulosa var. (Woolson 1878)
" sp.
" caryophylloides
" truncata (California)
" flavescens (Rky Mts)
" Chrysantha (" ")
" Canadensis (U.S.) Wild Columbine.
" Olympica (Parkman)
Delphinium cheilanthum (Regel March 22 1877)
" elatum var. [illegible]
" Maackianum (Regel March 15)

Bed No 5.

Delphinium consolida (Europe)
" Brunonianum (Kew)
Aconitum Orientale (Caucasus)
" Pyrenaicum (S. Europe)
" Californicum
" bicolor (Parkman)
Delphinium Lycoctonum (Europe)
Aconitum ferox (E. Indies)
" lasiostomum
" uncinatum Monkshood (U.S.)
Actaea alba White Baneberry. (U.S.)
" spicata var. rubra. (")

Bed No 6

Cimicifuga Americana
" racemosa.
Zanthorhiza apiifolia (Alleghanies.)
Paeonia tenuifolia (Siberia)
" officinalis
" anemonaeflora (Ware March 7 1879)
" sp. (May 15 1875)
" Montana (China)
" paradoxa (Regel)
" sp. Dr Hall. (Japan)
" albiflora. (Siberia)

Bed No 7

Caulophyllum thalictroides (U.S.)
Epimedium lilacinum
" alpinum (Europe)
" macranthum
" rubrum.
Dyphylleia cymosa. Umbrella...
Jeffersonia diphylla (U.S.)
Podophyllum peltatum Mandrake (U.S.)

Bed No 8

Papaver bracteatum (Persia
" pilosum (S. Europe)
" sp. Dr Regel March 18 1876)
" somniferum (Europe)
" orientale major.
" alpinum. (Europe)
" Rhoeas var. Hort.
" umbrosum (Caucasus)
" hybridum (S. Europe)
" dubium

Meconopsis cambrica (Paris) (March 1879)
Papaver alpinum var Pyrenaicum (Hort Cantab March 11 1879)

Bed No 10

Adlumia cirrhosa (U.S.)
Dicentra eximia (Thompson Jan 18. 1878)
" cucullaria (U.S.)
" eximia (Alleghanies)
" spectabilis (N. China) Bleeding heart)
Eschscholtzia Californica (California)

Bed No 11

Corydalis lutea (Europe)
" glauca (U.S.)

Bed No 12

Barbarea vulgaris (B.G. 1878)
" intermedia (Algeria)
" praecox (Europe)
Nasturtium Armoracia (") Horse radish.
Cheiranthus Cheiri (B.G. Sept 16 1878)

Bed No 13

Arabis hirsuta (Europe)
" laevigata (Dawson)
" lucida (Europe)
" sp. (Edinburgh. March 1878)
" Ludovici. Woolson Dec 13 1878
" verna (Paris)
" cordata (Woolson)
" Billardieri
" ciliata (Regel)
" alpina (different) (Regel)
" oxyota (Siberia)
" albida Europe
" borealis. (Paris)
" Dommarchii
" colorata (Woolson 1878)
" sp. (Dr Regel)

Bed No 14

Lunaria biennis (Europe)
Aubrietia deltoidea (S. Europe)
" olympica (Leichtlin Feb 11 1878)
" Bougainvillia
" Cyrei - Hort (Leichtlin Dec 1. 1878)
Vesicaria utriculata (S. Europe)
" sinuata (")

Alyssum diffusum (Asia Minor
" complectum
" Holboellii (Hort Cantab Feb 18. 1879)

Bed No 15

Alyssum tomentosum (Europe)
" maritimum (")
" mutabile (March 11. 1879)
Draba contorta (Rot Sep 16. 1878)
" aurea
" [illegible] (Arctic Am.)
" frigida (Europe)
" lasiocarpa R.B. Sep 17, 1878.
" ramosissima (S. States)
Hesperis matronalis var candidissima (Thompson March 25 1874)
" violacea (Regel)
" matronalis var. (Europe)
Erysimum rupestre (Armenia)
" repandum (B.B. Sept 17. 1878)
" cuspidatum
" Perowskianum (Afghanistan)
" aureum.
Sisymbrium strictissimum.

Bed No 16

Braya Tournefortii (Hort Cantab March 11 1879)
Lepidium affine
" alyssoides (Brandiger Feb 17. 1879)
" montanum (" Dec 22 1877)
" latifolium (Europe)
" graminifolium
" smithii
Isatis tinctoria (Europe) (B.B Sept 16 1878)
Vellineuma orientale
Iberis Tenoreana (S. Europe)
" ciliata (Caucasus)
" sempervirens (S. Europe)
" jucunda
" [illegible]
" coriacea (Parkman)
Crambe cordifolia (Caucasus)
Malcolmia maritima (Europe) Virginia Stock)
" africana. (S. Europe)
Sinapis laevigata (Hort Cantab March [illegible])
Raphanus niger (Paris)
Thlaspi alpestre (Woolson)

†

Bed No 17

Cleome pungens (S. America)
Polanisia graveolens (Hort Cantab March 6. 1879)
Reseda complicata.
" truncata.
" erecta. (Palermo, Feb 17. 1879.)
" phyteuma " July 18, (1878)
" lutea (Paris B.S., Feb 14. 1879)
" mediterranea (Hort Cantab Mch 5. 1879)
" bojeana (Paris, Mch 28. 1878)
" odorata (N. Africa) Mignonette.

Bed No 18

Viola striata (U.S.)
" cornuta (S. Europe)
" cucullata (U.S.)
" tricolor.
Cystus hirsutus (Regel.)
Helianthemum vulgare.
" sp.
" sp.
" sp.

Bed No 19.

Dianthus deltoides (Europe)
" caesius (")
" giganteus (Thompson Mch 25 1878)
" barbatus (Europe) Sweet William.
" latifolius "
" plumarius.
" atropurpureum.
" caucasicus.
" arenarius.
" cyrophylla.
Jacquinia pungens (Sonora)
Tunica saxifraga. (Europe)

Bed No 20

Silene Pennsylvanica (U.S.)
" nutans (Europe)
" Sawadskyi (Regel March 27 1878)
" mutabilis
" Italica (Europe)
" inflata (")
" caucasica (Caucasus)
" Tartarica
" maritima (Europe)

Silene Graeffei.
" longiciliata.
" pinacarpa.
Saponaria ocymoides. Europe.
" officinalis (")
Gypsophila altissima.
" acutifolia (Caucasus)
" scorzonerifolia.
" trichotoma.

Bed No 21.

Stellaria Holostea. (Europe)
Cerastium Biebersteinii. (Weise Mch 7. 1878)
" grandiflorum (Woolson Dec 14. 1878)
" oblongifolium (U.S.)
" arvense. (Europe)
" Boissieri (Thompson. Mch 25. 1878)
Agrostemma coronaria.
Lychnis fulgens (Europe)
" Chalcedonica. (Asia)
" viscaria (Europe)
" Chalcedonica var caucasica (Caucasus)
" Senno. Woolson 1878.

Bed No 22

Claytonia parvifolia (California)
Talinum patens (Mexico)
Portulaca sp. (Hort. Cantab.) mch 6. 1879)
" denticulata. (Hort Cantab mch 12 1879)
Arenaria graminifolia (Europe)
" serpyllifolia, Hort Cantab Mch. 11 1879
" gypsophyloides var. viscosa.
" macrophylla. (Woolson Oct. 1878)

Bed No 23

Althaea rosea. var Hort. Hollyhock.
" officinalis (Paris mch 25 1878.)
" narbonensis (" " 26 1878)
" Taurinensis
" digitata.
" sicifolia(?) Paris mch 26. 1878.)
" cannabina (Kew)
Lavatera Thuringiaca (N. Europe)
Sidalacea acutifolia Oct 10. 1878
Anoda hastata (Mexico)
Malva moschata (Europe)
" alcea.

Bed No 24

Malope malvoides (Algeria)
Napaea laevis
" dioica (U. States)
Callirhoe involucrata (Nebraska)
Malva borealis (Hort Cantab Feb 18. 1879)
" moschata
Sidalcea humilis
" acutifolia.

Bed No 25

Abutilon marmorata
" (Jean Verschaffelt)
" (Duke de Malakoff.)
" roseum superbum
" Darwinii Sept 24 1877.
Sida napaea (U.S.)

Bed No 26

Malvaviscus Drummondii (Cambridge. Jan 16 1879)
" arborea
" mollis.
" borealis
Modiola geranioides (Mass Mch 7 1879)

Bed No 27

Hibiscus sp.
" sp. Texas Mr E. Woolson
" sp. "
" californicus (California)
" militaris (S. States)
" moschatus (U.S.)

Bed No 28

Hibiscus coccineus (S. States)
" sp. (Japan B.W.B. June 20 1877)
" vitifolius (Nubia)
" manihot (Oct 10 1878)
" althea var grandiflorus Mexico. Ap. 16.
Linum grandiflorum (N. Africa)
" tenuie (Hort Cantab Mch 7 1879)
" asiaticum (" " " 11 ")
" perenne. (N. China, Perennial Flax)
" usitatissimum (Hort Cantab Mch 12. 1879)
" pumile (" " Feb 14 1879)
" angustifolium (" " Mch 4 1879)

Bed No 29

Geranium Pyrenaicum var alba. (Europe)
" Ibericum (Armenia)
" Endressi (Europe)
" Sanguineum (")
" pratense (")
" cinereum (Ware Mch 7 1879)

Bed No 30

Geranium Phaeum (Europe)
" maculatum (U.S) Wild Cranesbill.
Erodium cicutarium (C.S. Sargent Jan 16 1879)
" manescavi (Ware Mch 7 1879)

Bed No 31.

Geranium ———— ∞.

Bed No 32

Pelargonium peltatum
Tropaeolum ———.

Bed No 33.

Tropaeolum majus. Peru. Nasturtium

Bed No 34

Oxalis rubella (Paris Mch 26 1878)
" oregana (Woolson).
" rosea alba (Thompson Feb 17 1879)

Bed No 35.

Ruta macrophylla. (Paris Mch 26. 1878)
Zanthoxylum Americanum (U.S.)
" Carolinianum (S. States)
" schinifolium (Japan)
Ruta sp.
" graveolens Europe.
Phellodendron amurense (N.E. Asia)
Euonymus nanus (Caucasus)
" europaeus. Spindle tree (Europe)
" Japonicus var. (Japan)
" gracilis var (")
" Sieboldianus (")
" atropurpureus spindle tree (U.S.)
" Japonicus (Japan)
" americanus var obovatus (U.S)
Rhamnus lanceolata. U.S.
" catharticus Buckthorn Europe
" Purshianus Oregon.
" infectorius
" Frangula. (Europe)

Ceanothus Americanus (U.S)
Sapindus mukorossi. Japan)
Staphylea Bumaldi (")
" trifolia (U.S.)
Cedrela sinensis (N. China)
Acer rufinerva var albo-limbata (Japan)
" spicatum (U.S.) (Mountain Maple)
" polymorpha var dissectum (Japan)
" glabrum (Rky Mts)
" tataricum var ginnala (N.E. Asia)
" polymorpha (Japan)
" carpinifolia (Japan)
Xanthoceras sorbifolia (N. China)
Koelreuteria paniculata (")
Diospyros virginiana Persimmon. (U.S.)
Aesculus parviflora S. States
" flava var purpurascens. (U.S.)
" glabra Ohio Buckeye. (W. States)
Pinus Austriaca.
Vitis bipinnata (S. States)
" vulpina (" ")
" indivisa (" ")
" cordifolia (U.S)
" Labrusca. Fox grape (U.S.)
" aestivalis Summer grape. (")
" pentaphylla var Japan.
" " "
Prunus spinosa Sloe tree. (Europe)

Section II.

Yucca filamentosa (S. States)
Rhus
Pinus Austriaca (Austrian Pine Eu.)
Akebia quinata Japan)
Stauntonia hexaphylla (")
Calycanthus occidentalis. (California)
Nandina domestica Japan)
" " var tenuifolia (Hort.)
Berberis Thunbergii (Japan)
" Canadensis (S. States)
" vulgaris (Europe)
" Fendleri. (New Mexico)
" vulgaris var dulcis (Europe)

Berberis emarginata Siberia,
" cretica (Crete)
" sinensis (China)
" repens. Oregon.
" trifoliata Texas.
" aquifolium Oregon.

Section C.

Syringa vulgaris
Robinia viscosa. (S. States)
Helianthus sp.
Sophora Japonica.
Astilbe decandra (S. States)
Spiraea ulmaria.
Alnus ——.
Iris sibirica.
" stricta
" plicata.
Viola striata.
Aster Novae Angliae var roseus. (U.S.)
" Tardiflorus
Elymus sibiricus (Siberia).
Aster incisus " (")
Asimina triloba Papaw U.S.
Aster commixtus.
Rudbeckia laciniata. Cone flower. U.S.
Hemerocallis fulva. Day lily (Europe)
Achillea millefol. (")
Iris florentina (S. Europe)
" pallida
" germanica.
Smilacina stellata (U.S.)
Veronica Virginica (")
Liriodendron tulipifera Tulip tree U.S.
Aconitum variegatum (Europe)
Ulmus americana White Elm (N. Am.)
Acer saccharinum Sugar Maple.
Tilia Europaea. Linden (Eu)
Quercus rubra. Red Oak. (U.S)
Schizandra chinensis (China & Japan)
Magnolia grandiflora Laurel Mag. (S. States)
" sp. ((Japan) (Hogg)

Illicium anisatum (China & Japan)
" religiosum
Cornus sanguinea. (Europe)
Idesia polycarpa. (Japan)
Pittosporum Tobira. (China & Japan)
Tamarix africana (N. Africa)
" chinensis (N. China)
Hypericum ascyron. (Siberia)
" graveolens. (Alleghanies)
" pyramidatum (U.S.)
" patulum. (Japan)
" Kalmianum (Niagara Falls.)
" androsaemum (Europe.)
" sp.
Abutilon avicennae
Hibiscus africanus. (Africa)
Catalpa Bignonioides (U.S.)
Entelea arborescens. (Paris Mch 26. 1877)
Philadelphus Gordonianus (Oregon.)
" coronarius Japan.
Kitaibelia vitifolia (Europe)
Hydrophyllum Virginicum (U.S.)
Staphylea trifolia Bladdernuts. (U.S.)
Sassafras officinale (")
Diplopappus umbellatus. (")
Juglans cinerea. Butternut (")
Gossypium nankin. (Palermo. July 15. 1875)
" maritimum. (")
Aster Tradescanti (U.S.).
Eupatorium cannabinum
Convolvulus Cneorum. (Europe)
Heracleum giganteum.
Philadelphus hirsutus. (Alleghanies.)
" inodorus var grandiflorus
Ulmus montana Wych Elm. (Europe)
Ilex opaca. (U.S.)
" laevigata (")
" glabra. Ink berry. (U.S)
" perado. (Madeira Holly)
" Aquifolium fr. aurea. golden Holly.
Nemopanthes Canadensis. Mt. Holly. (U.S.)
Cornus florida. Dogwood. (N. Am.)
Rhus copallina (U.S.)

Rhus. venenata. poison dogwood. (U.S.)
" Osbeckii. (Japan)
" glabra var laciniata. Pennsylvania)
" — . (Dr Parry. Mexico July 16. 1878)
" aromatica. (U.S.)
" glabra. (U.S.)
Thermopsis montana. (C.S. Sargent. Jan. 1879)
Crataegus oxyacantha. Eng. Hawthorn.
Dicentra spectabilis and Helianthus roots.
Melilotus marginatus, Hort Cantab. Feb 16. 1879.
Medicago lupulina " " " 21 1879)
Melilotus alba. (Europe)
Lupinus arboreus. (1879.)
" arcuatus. (C.S. Sargent Jan 10. 1879)
" Nootkaensis (Regel. Feb 17. 1879)
Trifolium striatum (Hort Cantab. C.S. Feb 16. 1879.)
" agrarium (" " march 7 1879)
Trigonella ornithopodioides.
" coerulea. (Europe)
" polycerata (S. Eur.)
" cretica. (S ")
Cornus alternifolia (U.S.)
Trifolium ochroleucum (Bot. C.S. Feb 17. 1879.
Securigera coronilla (Hort. Cantab Mch 7. 1879)
Genista sagittata (Europe)
Laburnum alpinum
" vulgare var adami. (Hort.)
Genista sibirica.
" tinctoria Woodwax. (Europe)
" candicans var umbellata. (Regel Mch 12. 1879)
" virgata. (Madeira)
Cytisus nigricans (S. Europe)
" capitatus (Europe)
Viburnum Lantago. Sweet Viburnum (N. U.S.)
Rhus vernicifera. (Japan)
Cytisus supinus.
Rubus Balfourianus.
Amorpha fruticosa. (S. States)
" canescens Lead plant (N. States)
Magnolia acuminata. Cucumber tree. (U.S.)
Lupinus sp. (Hort Cantab March 11. 1879)
Laburnum —
Aster angustifolius (N. States)

Chrysanthemum sp.
Wistaria frutescens Amer. Wistaria (S. States)
Colutea arenata. Bladder senna.
" arborescens.
" orientalis
" Halepica.
" gonolobus biflorus (Hort Cantab. G.S. Feb 27. 1879)
— purpureus (" " " " 1879)
Colutea aurantiaca.
— ?.
Eupatorium ageratoides. W. Snakeroot. (U.S)
Spiraea lobata. (U.S.)
Quercus robur. var pedunculata. English oak. (Europe)
Silphium perfoliatum. Cupplant. (W. States)
Robinia hispida. Rose Acacia. (Virginia & South)
Ampelopsis quinquefolia. Virginia Creeper. (U.S)
Caragana arborescens. Siberian Peatree.
" microphylla,
" frutescens.
" spinosa (Siberia)
" frutescens var.
Calophaca Wolgarica. (Siberia)
Solanum Dulcamara.
Tradescantia Virginica (U.S.)
Lespedeza cuneata (2 G.S. Jul 15. 1878)
" bicolor. var. (N. China)
Spiraea opulifolia Ninebark. (U.S.)
Vitis cordifolia Frostgrape. (U.S. (Illinois))
Astragalus alopecuroides (Thompson. May 31. 1877)
Vicia sylvatica. Thompson.
" pseudocracca. (Italy)
" gigantea (Woolson. March 12 1879)
" atropurpurea. (Hort. Cantab. March 11. 1879)
Lathyrus pisiformis
" filiformis? (R.B.G. 1875)
Astragalus — ?
Orobus lathyroides
" vernus. (Europe)
Vitis aestivalis Summer grape. (U.S.)
Abies excelsa. Norway Spruce. (Europe)
Dalbergia frondosa. (N.W.H. July 2. 1878)
Sabbatia latifolia. (Mrs Bache April 29. 1878)
" speciosa (Kew. July 16. 1878.)

Poinsiana Gilliesiana (Lunt. 1877)
Halimodendron argenteum (Siberia)
Cassia calliantha (Paris 1878)
Poinsiana Regia. (Lunt & Brewer Feb 5. 1878)
" pulcherrima (" " " ")
Halimodendron grandiflorum.
Mimosa pudica.
~~Rhus typhina Staghorn Sumach. (U.S.)~~

Station D.
Bed No 1.

[illegible] (Paris [illegible])
[illegible] (U. States)
" [illegible] (Mexico)
[illegible] (Simpson, March 28 [illegible])
" [illegible] (H [illegible])
[illegible]
[illegible] Paris [illegible]

Bed No 2.

[illegible]
[illegible] (Paris [illegible])
" [illegible] (Simpson " 28 [illegible])
" sp. ([illegible])
"
" [illegible] ([illegible])
[illegible]

Bed No 3.

[illegible] [illegible])
" [illegible]
" [illegible] (Caracas)
[illegible] ([illegible] 1878)
" [illegible]

Bed No 4.

[illegible]
Rumex [illegible] (Europe)
" [illegible] [illegible]

Bed No 5

[illegible] U. States

[illegible] 1819)
" " " "
" " " ")

Bed No. 8.

Bed No 7

[illegible] ([illegible])
[illegible] ([illegible]) –

Bed No 11

([illegible])

Bed No 11

[illegible] Hagot. ([illegible])
[illegible] ([illegible])

Bed No 12

([illegible])
[illegible] ([illegible]).

Bed No 13

[illegible] ([illegible])
[illegible] (")

[illegible] ([illegible])

Bed No 14

[illegible]

[illegible]

[illegible] Hieracium [illegible]

[illegible] (N. S.)

[illegible] (N. S.)

[illegible] (N. S.)

[illegible] (N. S.)

[illegible] (N. S. & Europe)

[illegible] (N. S.)

[illegible] (N. S.)

[illegible] (Mechanics [illegible])

[illegible] (N. S.)

[illegible] (N. S. Europe)

[illegible]

menziesii (Moreau)
[illegible] hirtifiindens
—
(Chile)
[illegible]

[illegible] sis ([illegible] states)
[illegible] [illegible] Australia [illegible]
[illegible] [illegible] ([illegible])
[illegible] – ([illegible] Atlantic [illegible])
[illegible] ([illegible] states)
[illegible] ([illegible])
[illegible] ([illegible])
[illegible] –
[illegible] –
[illegible] ([illegible])
[illegible] (Hildebrand [illegible])
[illegible] ([illegible] States)

Parnassia [illegible]
" caroliniana -
Symphoricarpos [illegible]

Acorus calamus (N.S. & Europe)
" " variegata (" ")
Saxifraga pensylvanica (N.S.)
Rhexia virginica (N.S.)
Saxifraga [illegible] (Japan)
" rotundifolia (Europe)
Rhodea japonica (Japan)
Dianthus armeria (U. States + Europe)

[illegible] angustifolia (N.S.)
[illegible] (Alleghanies)
Myosotis palustris -
[illegible] -
Trichostema dichotomum (N.S.)

[illegible] formosa (Alleghanies)
Viola palustris (N. America + Europe)
Boykinia aconitifolia (Alleghanies)
Liatris odoratissima

Section E.

[illegible]
Thaspium aureum (U.S.)
[illegible] alba -
[illegible] pilosa (U. States)
" [illegible] " "
Convallaria majalis (Europe)
[illegible] Canadensis [illegible] (N. America)
Clethra alnifolia White Alder (U.S.)
Staphylea trifoliata. Bladdernut (U.S.)
Magnolia acuminata. Cucumber tree (
Fragaria vesca. Wild strawberry (U.S.)
Pinus sylvestris. Scotch Pine (Europe)
[illegible] serrata -
[illegible] - U. States
[illegible]

[illegible] [illegible] (U.S.)
pacificum (")
[illegible] (N. States
[illegible] Ataea. Canberry, (N. Hemisphere
Pennsylvanicum var. angustifolium (U.S.)
procumbens (White Mts. of Europe)
procumbens. Creeping Winter Green. (U.S.)
[illegible]
[illegible] sp (Regel)
[illegible] (U.S.)
[illegible] White Mts. [illegible]
[illegible] (Himalaya Mts)
[illegible] [illegible] Mts)
a. tiliifolia -
denticulata -
[illegible] (U.S.)
[illegible]
[illegible] (S. States
[illegible] (Europe)
[illegible] (U.S.)
ia [illegible]
[illegible] (Rocky Mts. [illegible])
[illegible] [illegible]
radicans (Japan)
[illegible] (U.S.)
[illegible] (S. States
[illegible] [illegible] (U.S.)

[illegible] [illegible]
[illegible] (Europe)
[illegible] canadensis (... West)
Ranunculus amplexicaulis (Europe)
Stellaria [illegible] (S. States)
Silene [illegible]
Lupinus [illegible] ([illegible] Mch 7 1879)
Ruscus aculeatus (S. France)
Potentilla pyrenaica
[illegible] [illegible] ([illegible] Mch 26 1879)
— — |

[illegible] [illegible]
[illegible] (B.N.C.) (Japan)
[illegible] — — —.
Pinus strobus
[illegible] [illegible] (1878)
[illegible] racemosa ([illegible] Feb 13 79)
Anemone [illegible] (Europe)
" cylindrica (N. States)
Potentilla rupestris (Seed)
[illegible] [illegible] (C. S. Sargent [illegible] 18[illegible]
Rumex (B.N.C.) (Japan)
[illegible] — ([illegible] Mch 5 1879)
[illegible] [illegible] (Seed)
— — — .

Euphorbia [illegible] (Plants)
Yucca angustifolia
[illegible] [illegible] (Siberia)
Lychnis viscaria (Europe)
—

[illegible] [illegible] (Mch 22 1879) [illegible]
[illegible] borealis ([illegible] Sept 11. 1[illegible])
[illegible] [illegible] ([illegible] & Europe)
Matricaria [illegible] (Seed)
Anemone sylvestris (Europe)
Achillea [illegible] (Chamberlain. Mch 21 1879)
Pycnanthemum incanum (N.S.)
" lanceolatum (")
[illegible] species
— [illegible] [illegible] (C. S. Sargent 7 15
[illegible] americanus (Plants)
[illegible] [illegible] [illegible] [illegible] Mch
[illegible] (Seed)

utai (S. Europe)
[illegible] uttuo (Spain)
[illegible] (U. States & [illegible])
[illegible] (Europe)
(U. States)
utiliformis. (Arizona)
[illegible]
[illegible]
[illegible] (U. States
[illegible] (Asia Minor)
[illegible] (Hudson. 1815)
[illegible]
[illegible] (Hub. - [illegible] Dec 16th 1878)
[illegible] parissus S. Europe
utriculosa (Europe)
utea - (Europe)

[illegible] (U.S.)
[illegible]
[illegible]
racoa (Hudson)
[illegible] (U. States)
[illegible] (U.S.)
[illegible] (Bulletin Feb 11 187[illegible]

Arenaria diffusa (Leichtlin [illegible] 1878)
Gentiana sp. [illegible] Hoser.
Dianthus [illegible] (America.)
Drosera [illegible] (U. States)
Dianthus cruentus (Asia Minor.
Dianthus Requienii (Corsica)
[illegible] [illegible] ([illegible] Mts)
Erysimum [illegible] (Turkestan)
Dianthus [illegible] (Europe)
Silene [illegible]. (Thompson)
Lychnis Lagascae. (Leichtlin Nov 16. 1878
Saxifraga [illegible] (Watson)
[illegible] grandiflora (Japan)
Silene [illegible] (March 4. 1878)
[illegible] sp. (Colorado)
Dianthus [illegible].
[illegible] — —. var. aspera.
[illegible] [illegible] (Asia Minor)
[illegible] alpestris. (S. Europe)
Silene Caucasica
Thalictrum minimum.
[illegible] obtusata (Thompson)
Dianthus fragrans (Regel)
[illegible] caespitosa (Thompson)
[illegible] Caucasicum (Caucasus
Cerastium grandiflorum (Watson)
[illegible] [illegible]. (Thompson)
Potentilla pulcherrima
—

Alchemilla [illegible].
Scabiosa [illegible]
Atragene alpina!

--

Fuchsia procumbens (New Zealand)
Cypripedium [illegible] (U. States,
[illegible] praecox. (U. States
Oxalis spectabilis
Lilium [illegible] (Siberia.
Cypripedium arietinum U. States

[illegible] repens. (Near Mich? 1879)
Sedum Nevii – (S. States
Lilium [illegible] (Siberia
Sedum [illegible] (Live forever)
Lilium Washingtonianum (California)
Spiraea caespitosa (Utah)
[illegible] Elliottii (S. States)
Chrysanthemum Tchihatchewii (Asia Minor)
[illegible] [illegible] –
" [illegible]sina –
Spigelia Marilandica Pink root – (U.S.)
Monardella odoratissima
Origanum umbellatum.
Veronica rupestris (Ture)
Scabra repens – (Caucasus)
Veronica fruticulosa (Europe)
Dianthus deltoides (")
Campanula pumila [illegible] (Near Mich? 1879)
" pulla –
" muralis –
" Hostii alba –
" rotundifolia
Veronica pectinata –
" agrestis –
[illegible] [illegible]sina – (E[illegible])

[illegible] [illegible] (U. [illegible])
[illegible] [illegible] ([illegible])
[illegible] [illegible] —
[illegible] [illegible] (California.)
Sedum telephioides (Alleghanies)
Trifolium minus — (Eu.)
Xerophyllum asphodeloides (U. States)
[illegible] — [illegible] [illegible] —
Sedum [illegible] (Japan)
[illegible] pratensum —
Campanula [illegible] (Regel)
Salix retusa (?)
" alpina —
" Drummondii Rky Mts
Astragalus tragacanthus (France March 1819)
Eriogonum (S. W. U. States. [illegible] Dec. 14. 18)
" ursinum —
" compositum —
Bulbine annua — (S. Africa)
" frutescens — (Palermo — July 16. 1878)
Allium stellatum (N. U. States)
" attenuifolium (California)
Moutan —
Allium tricoccum (U. S.) — — — —

Erythronium propullans — (U. S.)
" americanum (U. S.)
" grandiflorum (Rky Mts)
" Dens canis — (Eu.)
" purpurascens. (California)
" Dens canis purpureum (Eu.)
" albidum —
Fritillaria [illegible] — ([illegible])

[illegible] ([illegible])
[illegible] ([illegible])
[illegible] Scheuchzeri. (Eu.)
[illegible]um [illegible]! (Hot. [illegible] Feb 1879)
[illegible] junceum (Eu.)
[illegible] [illegible] ([illegible] 1879)
[illegible] (S. Pater)
[illegible] [illegible] (U.S.)
[illegible] cordifolia (Syria)
orientalis
latifolia (Eu)
[illegible] Fendleri. New Mexico)
[illegible] ([illegible])
[illegible]folia ([illegible] 1 1879.
[illegible] latifolium (Eu.)
— — [illegible] -
[illegible] -
Kamschatica -
[illegible] (Hot. [illegible] 1879)
[illegible] — — .
[illegible] -
[illegible] (N[illegible] Dec 14 1[illegible])
[illegible] rupestre -
[illegible] -
[illegible] (Eu)
[illegible]tteri (Bot Sept 16. 1878)
[illegible] -
Hartwelli -

[illegible]ea oppositifolia (U.S.)
[As]arum Canadense. Wild ginger – (U.S.)
" europaeum. (Eu)
Sisymbrium [illegible] –
Vaccinium Vitis idaea Cowberry.
Chiogenes hispidula (U. States)
Trillium erectum var. alba – (U.S.)
[illegible] tricuspidata. (Siberia)

[illegible] spathulata
Habenaria [illegible] (Vic. Hubbardston Mass)
Cypripedium [illegible] (U. States)
" occidentale
" acaule. (U.S.)
" venustum. (Nepal)
" candidum (U.S.)
" parviflorum (")
" spectabile (U. States)
" pubescens. (" ")
Habenaria Hookeri. (" ")
" hyperborea (" ")
" blephariglottis. (U.S.)
" tridentata
" fimbriata
" ciliaris
Goodyera pubescens. (E. States)
Pyrola rotundifolia
Epigaea repens. (May flower. States)

Hyacinthus candicans (Africa)
Commelyna Virginica (U.S.)
Hypoxis erecta. (")
Habenaria orbiculata (N. States)
[illegible] (U.S.)
Vaccinium erythrocarpum
Liparis liliifolia –
Orchis incarnata (Royle)
[illegible] –
Papyrus Antiquorum (Egypt)

www.ingramcontent.com/pod-product-compliance
Lightning Source LLC
Chambersburg PA
CBHW022016190326
41519CB00010B/1538
* 9 7 8 3 3 3 7 3 3 7 1 9 3 *